Pastoral Dreams

Dr. Sommer —
Thank you for your assistance.

Faithfully,
Anita Schnedl
10/22/77

Innocence, scissors-cutting, American, 1801.
Courtesy, Irvin G. Schorsch, Jr.

Pastoral Dreams

ANITA SCHORSCH

A Main Street Press Book

UNIVERSE BOOKS   New York

Library of Congress Catalog Card Number 77-70472
ISBN 0-87663-975-9

Published by Universe Books
381 Park Avenue South
New York City 10016

Produced by The Main Street Press
42 Main Street
Clinton, New Jersey 08809

Printed in the United States of America

Cover design by Quentin Fiore

The author gratefully acknowledges the cooperation of O. J. Rothrock, Curator of Graphic Arts, The Princeton University Library Department of Rare Books and Special Collections; Ruth Piwonka, Columbia County Historical Society; Beatrix Rumford, Director, Abby Aldrich Rockefeller Folk Art Collection; Dr. Frank Sommer, Librarian, The Henry Francis du Pont Winterthur Museum Library; staff of Beinecke Rare Book Library, Yale University, and of the Jenkintown (Pennsylvania) Library; the efforts of Doris Dinger; the photography of Nicholas Schorsch; and the patience and support of Irvin G. Schorsch, Jr.

Barnyard Scene, Mary Rees, needlework picture, American, 1827. Courtesy, Abby Aldrich Rockefeller Folk Art Collection.

Coverlet, patch and appliqué, cotton, American,
early nineteenth century. Courtesy, The Henry
Francis du Pont Winterthur Museum.

Fifteen years ago, in the face of a landscape increasingly urbanized and relentlessly mechanistic, I attempted something different. I raised a flock of sheep. And only ten miles distant from one of the world's largest cities. The result of that experience, a dividend far surpassing the merely temporal, is a lesson I shall not soon forget. For, of all the creatures on God's earth, sheep tell the story of our past, the history of our forebears, and the hope religious people everywhere share in a more peaceful life here and hereafter. The effort of preparing to survive each day was a text I read from the sheep in my pasture. Their collective history of 10,000 years of service seemed current and meaningful when I learned how to use the fleece off their backs, a new kind of work which slowed my day down to the turn of a wheel and the throw of a shuttle. I began to feel the labor of all primitive life, and the love of that labor, and the beauty of a docile animal, which, before my blinders had been removed, I had viewed as merely common.

I was the proprietary pastor of my flock, looking after the sheep as part of my family. They supplied me with a commodity, raw fleece, that most women did not (and still cannot) purchase at a neighborhood store. I, in turn, tried to supply my family with useful textiles: coverings for the body such as sweaters, hats, and hose; and material for the house to be made into curtains, bed rugs, and blankets. I would have enjoyed sharing the yarn and woven fabric with my neighbors, had any been left over. But, of course, as in traditional homesteads of the past, none was. Women then might have spun a little extra for the neighbors, but until they had more freedom and more time, they did not merchandise sheep, wool, or the woman-skills I now practice, such as spinning, weaving, and embroidering.

Sheep were once the most practical agricultural commodity, a walking seventeenth-century department store. From them, one derived food, fertilizer, warmth, and light in its many dimensions. Although I did not know how to use the horns, skins, or sheep's gut for making musical instruments as the ancients did, whenever I put tallow into a hanging crusie and lit a piece of scrap-worsted in the curved fat lamp support, I was reminded of Old Testament Hebrews rendering unto God a burnt offering. After counting the hours of concentrated effort it took me to shear, card, spin, wind, and wash the inch of scrap-yarn I used for a wick, I felt sinful burning it. I tried to contain my spinning to the daylight hours. At first I thought my time was more valuable than that of my ancestors, but the drops of fat, scraps of wool, and the hours of sunlight that measured my labor must have measured theirs, giving us all a sense of the divinity in work.

I became a good spinner, and life became slower and slower, but not much easier. It was fun to feel the rhythm of spinning once you made your foot press down in time to turn the wheel. Singing kept you in rhythm and in good humor. If your thumb and forefinger stretched the airy, oily woolen rolag in time with your foot pressing down, the wheel would tug it from between your fingers, and you could feel it coming out smooth thread all the way. You could test your own skills by following one labor with another, and if yesterday's spun yarn turned out to be today's lumpy sewing threads that stopped at the eye of a needle, you would remember to be a better spinner tomorrow. Sometimes the sin of error does more good than the grace of expertise.

The right sheep—greasy, kinky, and long-stapled—were not easy to find. But yeoman-farmers who started with the aptly described "Native Ewe, Wild, and Worthless . . . bred by and the Property of Nobody in Particular" had as much of a problem as I did finding suitable stock. Had I been a Quaker immigrant living anywhere along the eastern coast of North America from New England to the Barbados, I would have sent word to Rhode Island Quakers through my Yearly Meeting for a tight-fleeced, pregnant ewe. In those distant days, the Friends could depend on one another and on God's beautiful hills of the Narragansett Bay to protect flocks that roamed like Israel's children. Our fathers—Puritan, medieval, and Old Testament—saw abundance as divine blessing and a good flock as the finger of God identifying his faithful elect. My flock is small, their lesson great.

I breed white Cheviots that I can handle in a limited but green pasture. Though sheep need to walk to eat, I must not lose them to the neighbors' fields or to their dogs. Many feudal and frontier wars were fought because sheep turned the neighboring corn fields into dust. The seventeenth-century farmers kept common sheep—bare-bellied, long-legged, loose-fleeced—and requiring little care. The eighteenth century, however, saw the development of gentlemen-farmers in Europe and North America, men who thought more scientifically about their breeds and their pastures. Not satisfied with "bread and butter" sheep—that is, with merely putting mutton upon the table—experimenting farmers in the New World secured the heavily-protected and finely-fleeced European animals that made New England, and later the Southwest, a shepherd's dream. The battle of the breeds began with "smug, heavy, little Southdowns" and elegant, roll-collared Merinos.

My sheep have clean faces and tight, crimpy wool. They are the shade of early English cream-colored ware from Leeds. Natural white is not really white. It takes twentieth-century bleach to make a fleece completely white. But even nitre and urine, the first natural bleaches, could not remove the blemishes on Jacob's biblical sheep. The work of mordanting and boiling makes the creamy fleece take the dye of many colors. I know now, after such strenuous labor, why the phrase "dyed in the wool" means "to be thorough." Color was one of few, bright, happy changes in many early rural households, and nature was the source of the many variations. Like women of earlier generations, I boil oak bark or onion skins for yellow and madder root or walnut hulls for shades of brown. Rhododendron leaves make a grey that fades a bit too soon. If I want a deep, permanent blue, I send for indigo. And, I might add, the mails take almost as long now from New Zealand as they did in the eighteenth century from India. Black, the most exciting, contrasting color, is also the most undependable. It disintegrates or bleeds through every washing. It was expensive, too, in the eighteenth century, costing sometimes as much as a pair of silver shoe buckles. For my black dye, I found, as I guess my ancestors had, a naturally dark Karakul ram whose history went back to the Steppes of Asia. His wool is very black, and the staple very tough, as if he had been bred in a warm climate. He is beautiful and black, however, only for the first three months of his life, his fleece gradually fading, his crimp lengthening to a straggly, long fibre. In an effort to provide myself with subtle heather blends, the Karakul ram was allowed to tup my Cheviot ewes. In addition to the gift of color, the union produced a change in wooly texture. As I spin the blend, I can appreciate a new elasticity in the unyielding, hairy, black fleece.

Engraving from *The Panorama of Professions and Trades* by Edward Hazen (Philadelphia, 1837). Courtesy, The Library Company of Philadelphia.

My pastoral experiment produced a few surprises, not the least of which was Polka-Dots, a spotted lamb that looked like Jacob's scriptural sheep. Considering the moral implications of being blemished, Jacob must have been a very disappointed shepherd to have received Laban's ugly daughter and all the blemished sheep of the flock for his seven years of hard work. But I find my spotted ewe quite delightful and the mixed shades far easier to use than boiling dye or carding two colors together. And I am glad not to be living in a society that identifies goodness with being white, male, and unblemished. Farmers have never decided whether crossbreeding improves the breeds by mixing them or whether such hybridizing merely sacrifices vital individual qualities. But, like all men, they still dream of producing a pastoral El Dorado: the perfect sheep.

My work with sheep tires me, challenges me, satisfies me, constantly amazes me. I often watch two of the rams playing together. They both look ferocious with their large, dark horns curled around their ears. They stamp at each other, back away, and then charge. They are as powerful, as playful, and, as it were, as full of sensual desire as the cupids and psyches of mythological Greece. They suggest the fertility of spring, the love of life on earth. Their ewes remind me of Titian's paintings of two kinds of women, or perhaps the two natures of one woman: the sacred and the profane. From their profanity or earthiness—their fat, their grease, their warmth—I derive so much of the material sustenance I need. They require very little from me in return. The more they have, the more they give. I see why the ancient Greeks chose to show the Goddess of Plenty with a sheep, why a seventeenth-century emblem for taxation was a man with sheep and shears, and why America was depicted as the land of plenty with spinning wheels and sheep.

Sheep give their gifts in the first season—spring, the time of Venus, love, fecund showers, singing birds, lambs, and clammy, mucuousy, wonderful fleece. The shearing season is whenever the first real warm spell comes. English eighteenth-century mezzotints portraying the seasons show the shearing scene in June or July, which, I hope, does not fool one into thinking we all shear at that time. I shear in March. My sheep are always frightened of the stranger— the shepherd who comes into their pasture once a year and relieves them of the coat which has kept their pores from breathing all winter, and them from freezing. I do not blame the sheep for being afraid. The shearer is an outsider not of their flock. But he takes good care of them, running the shears over their skins without a nick, clipping their hooves, and worming them. And I know, from bringing new animals into the barn, that sheep can live harmoniously with all kinds of strangers. In fact, Puritan colonists in America like Edward

Shearing Time, engraving from *Rhymes and Roundelays in Praise of a Country Life* (New York, 1857). Courtesy, Princeton University Library, Department of Rare Books and Special Collections.

Johnson noted that such diversity was good: "Until the Land be often fed with other Cattell, Sheepe cannot live." The harmony of nature is best witnessed in the diverse creation of God, the like living with the unlike in mutual benevolence.

The shepherd shears my blemished lamb Polka-Dots and the weaker ones with special care. When his sheepdog nips too hard at their feet, the shearer catches the lambs with an old-fashioned crook. Like Jacob, he goes in with a staff and comes out with a flock. When he sits them up on their hindquarters, holding them skillfully with one hand, they grow quiet for him, much as a lamb without voice being led to the sacrifice. He shears them like a top being slowly spun. The only strenuous move left for him is to tie the rug of fleece that he takes off each back. He is one of the good shepherds. There are both kinds in this world.

Although sheep are a bit errant in their ways, and need a little guidance and forgiveness, I am not pressed to do any more shepherding for them than to put up a protective fence. I seed—I think unnecessarily—with a little red clover. I know that sheep, unlike other grazing animals, can eat scrub. I give them a covenant of salt, and I know they will let me move a little closer to them if, once in a while, I offer a handful of molasses and grain. Yeoman-farmers of old liked to give them turnip and potato treats. We put a little fat and a little heat to their bodies and, consequently, a little warmth in our hearts.

My pasture land has a free current of water running down the side of the hill. It quenches the sheep's thirst and keeps the pasture drained. The high, dry hill that early Puritan divines likened to the Scriptures prevents a hoof disease sometimes called ironically "the rott of heresie." Some days I feel that the mythic beauty at the top of the hill—the wispy clouds, the element of air, the summit and nearness of perfection—bears an estranged relationship to the earthy qualities of the dirt underfoot.

Hypocrisy, engraving, American, 1847. Courtesy, Sinclair Hamilton Collection, Princeton University Library.

I want to sweeten my animals' pasture. I use what was called "plaister" in the eighteenth century and what we now call lime. Plaister, popularized by a Philadelphia patron of agriculture named George Logan, was another result of the intellectual and scientific approach to farming of the day. Gentlemen-farmers, patronizing the useful as well as the noble arts, could join "those who labour in the earth and are the chosen people of God." Farming was a healthy employment that challenged a man's moral nature as well as his physical and mental condition. It was no mere coincidence that the church was called "a little-little flock of sheep," or that a gentleman was thought to have the sheep's frame of spirit. Men who pursued an experimental interest in farming, and sustained the expensive mistakes in feed and breed, shared the results with yeoman-farmers through essays published in farmers' journals from London to Baltimore. Despite a tentative distrust typical of men who learned to farm through "seat-of-the-pants experience," vocational farmers benefitted from the more rational approach to sheep raising demonstrated by the gentry. Next to the Bible, John Smith's *Memoirs on Wool* (1747), and the zodiac-moon charts, most farmers depended on American and English agricultural journals for the latest information on shepherdy.

I often walk down by the pasture to see my sheep. They do not always come up to the barn on their own accord unless it is very hot. Natural crevices, rocks, and bushes protect them. They die more easily from the pneumonia that comes with being overheated, or from starvation that follows the loss of their only half-set of teeth, than from the cold, unless it is one of those fierce snowfalls one finds in the upper regions of North America. But I like to know my ewes are in the barn during lambing season. My barn, which is built into the side of the hill the way old barns always have been, holds the warmth of manure in the dirt floor for the new babies. In spring, I always seem to be waiting for the moment when the ewes increase and multiply, anxious to help each ewe dry off her new baby, consciously a spectator glimpsing infinity. Each birth is like the manger scene over and over again, every lamb a new beginning, each one simple, unadorned, and naked without its fleece, more an essence than the material of this world, much like the naked babies John and Jesus playing innocently with the lamb and with each other.

Do you hear it? The lamb begins as a sound. It bleats for its own identity and that of its mother. Mother and child can find each other only when they accept their mutual sounds. A ewe will nurture no one else's baby, as if hers is the only one at God's right hand. Once in a great while, she will not feed even her own. Human mothers are like that too. But most of the time, when a new-born lamb stands wet and shaking in the center of the flock, bleating for direction, its mother returns the sound and ends the search. Every bleating baby seems to keep me and the sheep in a trance. People are not the only ones who talk, who need direction, and who seem to dream.

The Grave of William Penn, Edward Hicks, oil on canvas, Pennsylvania, 1847. Courtesy, The Newark Museum.

L'Agneau, engraving from *La Mère L'Oie*, French, mid-nineteenth century. Courtesy, Free Library of Philadelphia.

It is delightful and not in the least boring to watch them and be drawn into their dreams. Sometimes the sheep look lonely to me as they climb the hill. I lose sight of them when they go over the top with a seeming inner direction, bent on some wilderness pilgrimage. I wonder if sheep have a better sense of where they are going than the shepherds who lead them. I wonder, too, if I am projecting my own loneliness upon them—or merely anthropomorphizing.

Like one of Bunyan's pilgrims, some days I, too, climb the delectable mountain with them, remembering not to be too close and not to move too fast. If I frighten them, they will leap from scattered places in the pasture, plowing the ground with their cloven hoofs as they run together and then, huddling, behaving as if numbers will protect them. I used to think that they were watching over one another's souls. Confronted by imminent danger, possibly death, their only recourse is to run, and, beyond that, to stop, huddle, and accept death, if it comes, without a sound. The lamb has always been associated with death. Some say they are born with a desire to die, but I think they are born with the patience to live.

I often stand quietly and watch them stop their frantic running, suddenly, inexplicably. Then, like a beautiful dream, they fan out again. The sheep-path could lead to heaven, if by heaven we mean Arcadia filled with limitless time for contemplation, with flowers to perfume the air, and with trees to lean against. Idle? Perhaps. Who knows whether Little Boy Blue lying in the haystack, his eyelids closed, wasn't quietly composing lyrical poetry instead of snoozing. I am reminded of paintings by Gainsborough and Copley, of lovely

15

young ladies from both sides of the Atlantic posing in pastoral gardens, holding crooks with spontoons on the tips to protect the sheep, and caressing lambs that sit without struggle in their satin laps. The pose has to be a wishful dream, a dream to be a shepherdess like Rachel, or a saint like Agnes, the warm Flora, the ruling Faerie Queene, the wanton yet innocent bride of Miles Standish. My sheep would not sit still for such fussing as I observe in these eighteenth-century portraits. No sheep would. The lamb and the lady in these mannered portraits often share resemblances: the shape of their eyes and the bow of their mouths are often identical. They share, too, a common symbol—the Eternal Feminine—and a charitable nature.

From where I sit under the tree at noon, the whiteness of the barn has no shadows. The sheep enter the barn on one level and I on another. I keep my eighteenth-century looms and wheels upstairs where old-time farmers would have stored hay. A good craftsman could easily have built my barn, as he indeed fashioned my tools, and have been proud to have wrought both. The spinning wheel and the loom, thanks to the moral engravings of Hogarth and the city seal of Pastorius have come to represent the virtues of human toil. I think of threading the heddles and beating the weft threads, of throwing the shuttle and waxing the wool, of needing more hands and having no time to spare. Raising sheep, spinning, weaving—all are work, hard work, but pleasurable work—work that can be measured tangibly, measured in skeins, in miles walked or treadled, in bushels of barley. A hundred sheep were once equal to all the carts, plows, barrows, and rollers on a 200-acre farm or a week's work spinning thirty-six skeins of wool, labor that took almost a hundred miles of walking to spin.

A skilled cabinetmaker was proud to put his signature to the tools of the woolen trade. He often inventoried his wheels, clock winders, niddy noddies, and warping boards on the same ledger pages with his fine chairs, chests, coffins, and candle stands. When I spin I see the vase and ring turnings of the spokes going round with my wheel, casting exquisite shadows on the floor. Like the spindles of any fine Windsor chair, the wheel turnings are elegantly formed and proportioned. The mellow patina of the wood surfaces still retain the craftsman's loving care. The wheel is stamped with the marker's name, a pride in labor all too lost today.

A good carpenter also made the fences for the pasture. Anyone who has put a wood fence into the ground comes to value skill, sweat, and money. American cabinetmakers' account books meticulously list the cost of fencing labor. Three days of fencing in 1780, for example, was worth two coverlets or twenty-four yards of weaving, a good coffin, five pounds of sheep's wool, or a little less than "fore sheeps skins." Add a fourth day of fencing and a man's labor was worth almost as much as six "fraim chairs" or "four yards of chekt woolen," then infinitely more valuable than mere plain cloth. Spinning two spindles and a half of fine yarn was worth even more than the four "chekt" yards.

Although men and boys rarely spun, they did weave. Itinerants carried two- or four-poster portable looms that fit together easily because the craftsmen marked them with numerals in each corner, like unassembled toys for today's children. Such weavers went from house to house weaving the threads that had been prepared by the women. Like a traveling portrait painter, an itinerant weaver lived a week or two or three with a family until he had woven what the family could afford in thread and cash.

Trade card, engraving, English, eighteenth century. Private collection.

To many an early family, the quality of a man's work was not only the measure of his success but the measure of his faith. At least it was one measure of God's presence in the Protestant yeoman. Quality was communicated not only in the tools of the tradesman, but in the symbol of the lamb itself. In America, from the pewter mark of William Will and the clothing of Brooks Clothiers (better known as Brooks Brothers) to the seal of Germantown, Pennsylvania, and that of Jamestown, Rhode Island, wool was a fiber that assured

17

men of quality. I once wanted to call my farm The Woolpack, The Staple Farm, or even The Holy Blaise. It would not have hurt me one bit to have had the patron saint of woolcombers over my door. I could have hung a ram and a teazle emblem outside the door as had the old English public-houses if I were not wary of my own ingenuousness and afraid that people would regard my farmhouse as a quality motel. Pennsylvania, in fact, once had six lamb taverns. And there was one in Baltimore, one in Boston, and one in Ohio. They called the Boston inn *Sheep's Baby* and used it as a counting house, exchange office, reading room, and bank. No hospitaler of the Crusades could have made both merchant and farmer feel more comfortable or more welcome than did the sign of the *Sheep's Baby*. The eighteenth-century innkeeper was more often than not the same farmer who owned the sheep in the barn next to the tavern—a woolen mill proprietor, a town counsellor, a deacon in the church, and an influential man in the community. The tavern keeper was not troubled by modern ideas of specialization.

Tin toy, American, mid-nineteenth century.
Private collection.

Some days I sit and sketch the sheep as they lean against the railed fences and scratch their backs. Some of their woolen fuzz catches on the ragged wood, and neighboring children collect it and take it home to their mothers. It makes their pockets and their fingers very greasy. Modern mothers do not always like that.

I have many fence watchers. They like the sheep. I, too, am pleased with my flock, perhaps proud, in my daydreams not unlike the Albany banker Elkanah Watson, or the farmer-merchant Robert R. Livingston of New York, or David Humphreys, protégé of General Washington and a poet, an interesting one. These and many other wool breeders in North America were, like the ministers of their churches, hopeful that the success of their flocks was a sign of good faith and the coming kingdom. Wanting to purify and elevate sheep breeds, these men secured prize Merino rams, the pride and power of Europe. During the subsequent "Merino craze," each family had to have two Merinos in every barn, a sure symbol of status. Delighted with the actual quality of the wool, men merchandised the Merino by claiming it was a good capital investment. The higher the percentage of Merino strain bred into the flock, the more poundage and pennies would be in the fleece. The stately beauty of the animal was promoted as a sign of the breeder's good taste. Men wanted for their flocks what they wanted for their families. They commissioned portraits to be painted of specimen sheep, and artists, consequently, placed ads in farm journals to make themselves and their talents readily available to the new social trends in husbandry. It took a financial panic in the 1830s to turn a less than realistic enthusiasm for this fine-fleeced creature into a financial disaster and to reduce Merinos from one-thousand dollars a ram to less than a dollar a head.

19

Ram weather vane, copper body and zinc head, c. 1860-65, American. Courtesy, The Newark Museum.

Chalkware sheep with girl, nineteenth century,
American, from the Index of American Design.

Early Protestant ministers believed that the universe could be seen as a woven fabric, a harmony of threads, the result of all good weaving and all good actions. In such a divine scheme, man was considered "the Lord's Yarn, the many threads of righteous Actions. . . ." I used to dream that all the world could live together and never fight, everybody "a baa lamb," as Louisa May Alcott used to say. Meekness, mildness, softness, innocence. But this is a child's dream, this playfulness—chasing the sheep, tackling the lamb, falling in the mud, stamping feet, crying, being comforted—all the sensual experiences that disrupt and yet please the innocent child in all of us. Harmony, on the contrary, divine order, unity, is an *adult* dream—not the product of childish innocence, but the result of intellect, wisdom, hard work, labor. The dream need not be abandoned. Harmony, living together in peace and love, requires the same work and concentration, planning and design, as weaving. True innocence is that complex.

Perhaps the lamb is not altogether a child's symbol after all. Even "Mary Had a Little Lamb" has a subtle adult moral at the end. The answer to the question "Why does the lamb love Mary so?" is simply that "Mary loves the lamb, you know," a recognition of mutuality, of equality, of the harmony inherent in a young democracy. It takes a long time to learn who loves us and a bit longer yet to know whom we love. The romance of the nineteenth-century pastoral was the love of nature, of women for their children, fathers for their sons, of brothers for brothers, all God's creatures large and small, weak and strong. The life, art, and literature of the time reflected the concern people had for others, and put this spirit to work destroying the sins of slavery, poverty, ignorance, inequality.

I am not the only spinner in the barn. Spiders work in every corner, spinning and weaving at the same time, a singular process, an extraordinary demonstration of natural technology. If only *we* could do that—thread and warp, weave and throw, spinning and weaving harmoniously. The center could then hold.

I sit on the hill under the shade of a tree. The sun is high now, but then all God's work and mine is done in the morning. We both have our joy in the noon day. I watch my sheep sit under a tree. The world is a big pasture. There is a tree for the sheep and a tree for me. There are many different kinds of us, but, like Edward Hicks's universe of animals, only the lamb sits in all four corners of the kingdom in a revealed light. The sheep need me sometimes, and I need them. But no longer for survival—mutton is unpopular; wool has been replaced by synthetics; horns are plastic; strings are nylon; parchment is wood pulp; and no one has yet figured out how to make the protein in a sheep's fleece digestible. Sheep represent not the survival of man's flesh, but of his spirit. We need a steady light of hope ahead, a sheep-path to follow to the top of the mountain, a

Needlework picture, Mary Upelbe, Massachusetts,
1767. Courtesy, Historic Deerfield, Inc.

friend to follow who is strong and predictable, innocent and pure, yet still of this earth, warm and soft, lively and fruitful. Am I looking for El Dorado?

I like to lie in the shade and read from books written in a day when shepherds and weavers were commonly encountered in the countryside, at a time when the simple evoked wonder and thoughtful words. Pastoral literature was meant for people who value sheep, who respect the earth and all its fruit, who find joy in labor and solace in the noonday rest—for those who love the spring, new babies, spinning—all these innocent beginnings that glisten in beautiful isolation.

At the start of the nineteenth century, William Blake, poet, artist, and mystic, penned the following famous lines, an apt conclusion to my pastoral dream:

> Little Lamb, who made thee?
> Dost thou know who made thee?
> Gave thee life, and bid thee feed,
> By the stream and o'er the mead;
> Gave thee clothing of delight,
> Softest clothing, woolly, bright;
> Gave thee such a tender voice.
> Making all the vales rejoice?
> Little Lamb, who made thee?
> Dost thou know who made thee?
>
> Little Lamb, I'll tell thee,
> Little Lamb, I'll tell thee:
> He is called by thy name,
> For He calls Himself a Lamb,
> He is meek, and He is mild;
> He became a little child.
> I a child, and thou a Lamb,
> We are called by His name.
> Little Lamb, God bless thee!
> Little Lamb, God bless thee!

Spring

It was now early spring—the time of going to grass with the sheep, when they have the first feed of the meadows, before these are laid up for mowing. . . . The middle of spring had come abruptly—almost without a beginning.

The landscape, even to the leanest pasture, was all health and colour. Every green was young, every pore was open, and every stalk was swollen with racing currents of juice. God was palpably present in the country, and the devil had gone with the world to town.

Thomas Hardy
Far from the Madding Crowd (1874)

American Homestead Spring, Currier and Ives, lithograph, New York, dated 1869. Courtesy, Museum of the City of New York, Harry T. Peters Collection.

The Dreamland Sheep: A Charm

When, tossing on your restless bed,
You can not fall asleep,
Just resolutely close your eyes,—
See a field-path before you rise,
And call the dreamland sheep.

They come, they come, a hurrying crowd,
Swift-bounding, one by one;
They reach the wall in eager chase;
The leader finds the lowest place;
They cross, and on they run.

Oh! many times on sleepless nights
I watch the endless throng,
Their pretty heads, their woolly backs,—
As crowding in each other's tracks
They press and race along.

I try to count them, but, each time,
Lose reckoning at the wall.
They come from where the gray mists blend,—
In mist they vanish at the end,
With far, faint bleat and call.

Off drop the day-time cares. Away
The nervous fancies fall;
And peacefully I fall asleep,
Watching the pretty dreamland sheep
Crowd through the dreamland wall.

Mary L. B. Branch
St. Nicholas (September, 1885)

Fantasy, oil on velvet, American, c. 1840.
Courtesy, Abby Aldrich Rockefeller Folk Art
Collection.

Mary Had a Little Lamb

Mary had a little lamb,
Its fleece was white as snow;
And everywhere that Mary went
The lamb was sure to go.

It followed her to school one day,
That was against the rule;
It made the children laugh and play
To see a lamb at school.

And so the teacher turned it out,
But still it lingered near;
And waited patiently about
'Til Mary did appear.

"Why does the lamb love Mary so?"
The eager children cry;
"Why, Mary loves the lamb, you know,"
The teacher did reply.

Sarah Josepha Hale (1830)

Mary and Her Little Lamb, oil on canvas, New England, c. 1830.

The Lady of the Lambs

She walks—the lady of my delight—
A shepherdess of sheep.
Her flocks are thoughts. She keeps them white;
She guards them from the steep.
She feeds them on the fragrant height,
And folds them in for sleep.

She roams maternal hills and bright,
Dark valleys safe and deep.
Her dreams are innocent at night;
The chastest stars may peep.
She walks—the lady of my delight—
A shepherdess of sheep.

She holds her little thoughts in sight,
Though gay they run and leap.
She is so circumspect and right;
She has her soul to keep.
She walks—the lady of my delight—
A shepherdess of sheep.

Alice Meynell (c. 1897)

Mary and Her Lamb, inscribed Laura Reed, silk
and paint on satin, American, c. 1800. Courtesy,
The New-York Historical Society.

The Live Murmur
of a Summer's Day

Go, for they call you, Shepherd, from the hill;
  Go, Shepherd, and untie the wattled cotes:
No longer leave thy wistful flock unfed,
Nor let thy bawling fellows rack their throats,
    Nor the cropp'd grasses shoot another head.
      But when the fields are still,
And the tired men and dogs all gone to rest,
    And only the white sheep are sometimes seen
    Cross and recross the strips of moon-blanch'd green;
Come, Shepherd, and again begin the quest.

Here, where the reaper was at work of late,
    In this high field's dark corner, where he leaves
      His coat, his basket, and his earthen cruise,
    And in the sun all morning binds the sheaves,
      Then here, at noon, comes back his stores to use;
        Here will I set and wait,
While to my ear from uplands far away
    The bleating of the folded flocks is borne,
    With distant cries of reapers in the corn—
All the live murmur of a summer's day.

Matthew Arnold
from "The Scholar-Gipsy" (1853)

The Cottagers, M. J. Dalzel, watercolor on paper,
American, c. 1800. Courtesy, Collection of The
Newark Museum.

The Two Sisters

Emma, being the eldest sister, had many things the baby could not have; but Emma, being a good girl, gave her little sister leave to look at all her toys, and you can not think how pretty it was to see Emma lead her little sister by the hand and show her all her things. She used to say, "See here, dear little sister, look at this pretty thing; do not break it, my love; if you do, I shall not have another like it, and then I know you will be very sorry. . . ."

Sometimes the little girl would keep Emma's toys too long, and Emma would want them: but she did not snatch them, saying instead, "I must teach her to be good and kind, by being good and kind myself, and mama will love us more and more, every day. . . ." All little children will find that it is much better for them to be obedient to their parents, and affectionate to one another; for they will not only be far happier, by so doing, but find themselves rewarded when they least expect it.

The Little Sisters: or Emma and Caroline (1840)

The Burnish Sisters, William Matthew Prior,
oil on canvas, American, dated on back 1854.
Courtesy, Edgar William and Bernice Chrysler
Garbisch.

Drink, Pretty Creature, Drink!

The dew was falling fast,
    The stars began to blink;
I heard a voice: it said,
    "Drink, pretty creature, drink!"

The lamb while from her hand
    He thus his supper took,
Seemed to feast with head and ears
    And his tail with pleasure shook.

"Drink pretty creature, drink!"
    She said in such a tone,
That I almost received
    Her heart into my own.

Parley's Magazine (March, 1833)

Pastoral Scene of Woman and Sheep, silk
chenile and paint on silk, American, c. 1790.
Courtesy, The New-York Historical Society.

The Sheep Has Its Enemies

Such is the mildness and gentleness, and patience of the sheep, that it has been in almost all ages the emblem of these virtues in man. In the spring, when the fields are spread with their green flowery carpet, what can be more delightful than to see a dozen or a score of lambs with their snowy fleeces, chasing one another round in circles, leaping up and down the rocks, and putting themselves into all manner of antic postures, manifesting the height of animal enjoyment! They afford much diversion for children, and give them much delight.

But the sheep has its enemies. Among the most destructive of these is the wolf; though sometimes the ram, with his hard forehead, will make a severe attack upon the dog, yet sheep in general are but poorly able to defend themselves against their many invaders. They depend upon man; to him in some sense they look up for protection; and seeing how much comfort he derives from them, he seems to be under obligation to make them as safe and comfortable as may be, the little time it is consistent for them to live.

Jonathan Fisher
Scripture Animals (1834)

The Shepherd and His Flock, watercolor on
paper, American, c. 1820. Courtesy, Abby
Aldrich Rockefeller Folk Art Collection.

If You Will Love Me

O my lambs! if you will love me who first loved you, and if you will love one another, I will take you hereafter to a fairer pasture than this. Look to the top of that mountain, which lies exactly before you, and you will see the skirts of a fair and lovely country. The sun is shining upon the trees thereof so as to make them look like the finest gold; beyond those trees lies a pasture-ground, in which I find delight to walk, and in which my sheep and lambs find perfect rest. To this place will I take you at some future time, if you will know my voice and hearken unto it.

The Two Lambs: An Allegorical History (1832)

The Christ Child Riding on the Lamb, William
Blake, tempera on canvas, English, dated 1800.
Courtesy, Crown Copyright: Victoria and Albert
Museum.

The Innocence of Childhood

When they came out into the garden, the child lifted her small hands toward heaven and exlaimed, "Only see once, mother, the sheep in heaven! a whole flock! How beautiful and how tender!"

They were, however, nothing but soft flakes of clouds. They appeared like a flock of lambs wandering over the pasture, white and curly, and they shone in the beams of the beautiful full moon. And the mother of the child lifted her countenance and beheld the clouds with mournful joy. For she remembered another feature of child-like simplicity, which draws down in the small compass of its thoughts the high splendor of celestial things and clothes them with terrestrial form and beauty.

This little girl saw in the clouds of heaven the sheep of the earth. "Happy art thou, O tender innocence of childhood!" said the mother, and pressed her daughter to her bosom.

Parley's Magazine (March, 1833)

Mourning Picture, Edwin Romanzo Elmer, oil on canvas, Massachusetts, c. 1889. Courtesy, Smith College Museum of Art.

The Peaceable Kingdom

The wolf also shall dwell with the lamb, and the leopard shall lie down with the kid; and the calf and the young lion and the fatling together; and a little child shall lead them.

And the cow and the bear shall feed; their young ones shall lie down together; and the lion shall eat straw like the ox.

And the suckling child shall play on the hole of the asp, and the weaned child shall put his hand on the cockatrice's den.

They shall not hurt nor destroy in all my holy mountain: for the earth shall be full of the knowledge of the Lord, as the waters cover the sea.

Isaiah 11: 6-9

Pastoral Landscape, Edward Hicks, oil on wood
panel, Pennsylvania, c. 1846. Courtesy, Abby
Aldrich Rockefeller Folk Art Collection.

Leisure

What is this life, if, full of care,
We have no time to stand and stare?

No time to stand beneath the boughs
And stare as long as sheep or cows.

A poor life this, if, full of care,
We have no time to stand and stare.

No time to see, when woods we pass,
Where squirrels hide their nuts in grass.

No time to see, in broad daylight,
Streams full of stars, like stars at night.

No time to turn at Beauty's glance,
And watch her feet, how they can dance.

No time to wait till her mouth can
Enrich that smile her eyes began.

A poor life this, if, full of care,
We have no time to stand and stare.

William Henry Davies (c. 1900)

The Shepherd's Grove, inscribed Agnes Yeakle,
silk on linen, Pennsylvania, dated 1811.
Courtesy, Edgar and Charlotte Sittig.

Little Bo-peep

Little Bo-peep has lost her sheep
And can't tell where to find them.
Leave them alone and they'll come home
And bring their tails behind them.

Little Bo-peep fell fast asleep
And dreamed she heard them bleating.
But when she awoke, she found it a joke
For they were still a fleeting.

Then she took up her little crook,
Determined for to find them.
She found them indeed, but it made her heart bleed,
For they'd left all their tails behind 'em!

Mother Goose Nursery Rhyme

Little Bo-peep, George Romney, oil on canvas, English, c. 1780. Courtesy, The Philadelphia Museum of Art: The John H. McFadden Collection.

The Domestication of Sheep

That we are not at this moment fierce, savage, and brutal, little superior to the beasts that roam in the wilderness is probably owing to the domestication of grain-feeding animals, and, first of all, to that of sheep. To them we are also indebted for some of the most pleasing, as well as for the most important and useful arts. The cradle of music and poetry was rocked by the shepherds of Arcadia; while the spindle and the distaff, the wheel and the loom, originated in the domestication of sheep.

This little animal, then, in losing its own wild nature, has not only converted the savage into the man, but has led him from one state of civilization to another; the fierce hunter it has changed into the mild shepherd, and the untutored shepherd into the more polished manufacturer. The more sedentary men became, the greater were their wants—and from that dependence originated civilization and polished societies.

Robert R. Livingston
Essay on Sheep (1809)

Mr. and Mrs. Sherman Griswold: Salting Sheep, attributed to James Johnson, oil on canvas, New England, c. 1831. Courtesy, The Columbia County Historical Society, Kinderhook, New York, Gift of Mr. and Mrs. Charles L. Rundell in memory of Mrs. Frank Rundell.

The Sheep Washing

Before shearing, the sheep undergo the operation of washing, in order to free the wool from the foulness it has contracted.

Upon the brim
Of a clear river, gently drive the flock,
And plunge them one by one into the flood;
Plung'd in the flood, not long the struggler sinks,
With his white flakes, that glisten through the tide;
The sturdy rustic, in the middle wave,
Awaits to seize him rising: one arm bears
His lifted head above the limpid stream,
While the full clammy fleece the other laves,
Around laborious with repeated toil;
And then resigns him to the sunny bank,
Where, bleating loud, he shakes his dripping locks.

(John Dyer, 1757)

The shearing itself is conducted with a degree of ceremony and rural dignity; and is a kind of festival, as well as a piece of labour.

"Sixth-Month"
Calendar of Nature (1815)

Sheep Washing, Benjamin West, oil on slate
panel, English, 1795. Courtesy, Rutgers
University Fine Arts Collection.

The Manners of Sheep

All up and down the greeny grass
The sheep in flocks together pass;
With nibbling noses hills are sown
And where they go the sod is mown.

With thick-set tails a-wag behind—
They roam or nibble with one mind:
And if one lifts his head on high
All other heads at once up fly.

As stones in field, then stand they still
Or run they all with single will;
And whether there is aught to leap,
All jump if jump the leader sheep.

Where learned the simple sheep such ways
No one had told in ancient days;
But now some think they learned them when
The silly sheep were silly men.

John Albee
St. Nicholas (May, 1891)

54

Hooked Rug, cotton and wool on burlap,
probably Pennsylvania, dated 1897. Courtesy,
Dr. and Mrs. Donald Herr.

A Lamb Is Innocence

Every natural fact
is a symbol of some spiritual fact.
An enraged man is a lion,
a cunning man is a fox,
a firm man is a rock,
a learned man is a torch.
A lamb is innocence.

Ralph Waldo Emerson
"Nature" (1844)

Mary Beekman with Lamb, John Durand, oil on canvas, New York, c. 1766. Courtesy, The New-York Historical Society, Gift of the Beekman Family Association.

The Passionate Shepherd to His Love

Come live with me and be my love,
And we will all the pleasures prove,
That valleys, groves, hills and fields,
Woods or steepy mountains yields.

And we will sit upon the rocks,
Seeing the shepherds feed their flocks
By shallow rivers to whose falls
Melodious birds sing madrigals.

And I will make thee beds of roses,
And a thousand fragrant posies;
A cap of flowers and a kirtle
Embroidered all with leaves of myrtle;

A gown made of the finest wool,
Which from our pretty lambs we pull;
Fair-lined slippers for the cold,
With buckles of the purest gold;

A belt of straw and ivy buds,
With coral clasps and amber studs;
And if these pleasures may thee move,
Come live with me and be my Love.

The shepherd swains shall dance and sing
For thy delight each May morning;
If these delights thy mind may move,
Then live with me and be my Love.

Christopher Marlowe (c. 1585)

Morning, attributed to Sophia Burpee,
watercolor on paper, American, c. 1810.
Courtesy, Abby Aldrich Rockefeller Folk Art
Collection.

Merino Sheep

The excellence of the Merinos consists in the unexampled fineness and felting property of their wool, and in the weight of it yielded by each individual sheep: the closeness of that wool, and the luxuriance of the yolk, which enables them to support extremes of cold and wet as well as any other breed; the easiness with which they adapt themselves to every change of climate, and yet thrive and retain, with common care, their fineness of wool: an appetite which renders them apparently satisfied with the coarsest food; a quietness and patience into whatever pasture they are turned, and a gentleness and tractableness not excelled by any other breed.

L. A. Morrell
The American Shepherd (1863)

Group of Merino Sheep, N. W. Wineland,
pencil on paper, Ohio, 1879. Courtesy, Museum
of Fine Arts, Boston, Karolik Collection.

Woman-Skills

Solomon thus describes the good wife:—"She seeketh wool and flax, and worketh willingly with her hands."

In every country where the simplicity of manners and virtues of the female are uncontaminated, spinning and weaving are the ordinary chosen employments. "She maketh herself coverings of tapestry; her candle goeth not out by night. She layeth her hands to the spindle, and her hands hold the distaff."

L. A. Morrell
The American Shepherd (1863)

Needlework Picture, wool and silk on linen, American or English, c. 1760. Courtesy, The William Trent House.

The Lesson of Sheep

"I wish," said the child, as it passed a flock of sheep and lambs, lying on the greensward,—"I wish I was as harmless as those creatures are. They have told no wrong stories, nor engaged in any cruel sports, nor have they ever been ill-natured or passionate, or revengeful. They do not appear guilty or ashamed when they look at each other, or at me, as I pass. But I, who am apt to think myself far better than they, can scarcely look at my superiors, or, above all, upwards towards Heaven, without feeling guilty. They—happy creatures—have no conscience!"

Parley's Magazine (1834)

Julia and the Pet Lamb; or Good Temper and Compassion Rewarded, engraving from book of same name, Portland, Maine, 1827. Courtesy, Sinclair Hamilton Collection, Princeton University Library.

Pastoral Scene

We get a peep at a meadow where sheep are lying, poor harmless quiet creatures; how still they are. Some lying socially side by side; some grouped in three's and four's; some quite apart. There are pretty lambs among them nestled in by their mothers. Soft quiet sleeping things. There is a party of these young lambs wide awake as the heart can desire; half a dozen of them are playing together, frisking, dancing, leaping, butting and crying in the young voice which is so pretty and diminutive of their full grown bleat. How beautiful they are, with their innocent spotted faces, mottled feet, long curly tails, light flexible forms frolicking like kittens with gentleness and assurance of sweetness and innocence which no kitten or cat can have. How complete and perfect is their enjoyment of existence!

Little rogues! Your play has been too noisy, you have awakened your mamas—two or three of the old ewes are getting up. One marching gravely to the troop of lambs has selected her own, given her a gentle butt and trotted off, the poor rebuked lamb following meekly but every now and then casting a look at its playmates. They resumed their gambols while the stately dam looked back to see if her little one was following. At last she lay down and the lamb by her side. It was the prettiest pastoral scene I ever saw in my life.

Juvenill Miscellany (1828)

Girl with Five Sheep, embroidered "Mary Ann
Rookin's Work at M. S. Evanes School, Penruh,"
wool, silk, and paint on silk, English, c. 1810.
Courtesy, Bertrand and Charlotte Rowland.

The Lamb

See the young lambs, so brisk and gay,
   With fleeces clean and white,
Sport o'er the meads in harmless play—
   How pleasant is the sight!

What mildness in their look appears;
   What innocence and peace;
Their sport the vernal season cheers,
   And makes dull sorrow cease.

But oft the lamb, when more mature,
   Must fall beneath the knife,
A greater blessing to secure,
   To save the shepherd's life.

Thus once the Lamb of God was slain,
   His life he freely gave,
From sin, and guilt, and endless pain,
   The sons of men to save.

What endless thanks we mortals owe
   For this display of grace;
Shall we reject the favor? No!
   Today the gift embrace.

Jonathan Fisher
Scripture Animals (1834)

Fraktur, from *Paradisisches Wunder Spiel*, watercolor and ink on paper, Ephrata, Pennsylvania, 1754. Courtesy, The Henry Francis du Pont Winterthur Museum, Joseph Downs Manuscript Collection.

The Sheep

Little sheep, pray tell me why,
In the pleasant fields you lie,
Eating grass and daisies white,
From the morning till the night?
Ev'ry thing can something do,
But what kind of use are you?

Nay, my little master, nay,
Do not serve me so I pray;
Don't you see the wool that grows
On my back, to make you clothes?
Cold, and very cold you'd get,
If I did not give you it.

True, it seems a pleasant thing
To nip the daisies in the spring;
But many chilly nights I pass
On the cold and dewy grass;
Or pick a scanty dinner where
All the common's brown and bare.

Then the farmer comes at last,
When the merry spring is past,
And cuts my wooly coat away,
To warm you in the winter's day.
Little master, this is why
In the pleasant fields I lie.

The Juvenile Casket; containing Short Poems,
Adapted to the Capacities of Young Children (1839)

Needlework Picture, wool, silk, metallic yarn on linen, Massachusetts, dated 1748. Courtesy, The Henry Francis du Pont Winterthur Museum.